WILDLY CUTE ANIMALS!

FENNEC FOXES

BY KRISTEN RAJCZAK NELSON

Gareth Stevens PUBLISHING

Please visit our website, www.garethstevens.com. For a free color catalog of all our high-quality books, call toll free 1-800-542-2595 or fax 1-877-542-2596.

Cataloging-in-Publication Data
Names: Rajczak Nelson, Kristen.
Title: Fennec foxes / Kristen Rajczak Nelson.
Description: Buffalo, NY : Gareth Stevens Publishing, 2026. | Series: Wildly cute animals! | Includes glossary and index.
Identifiers: ISBN 9781482471175 (pbk.) | ISBN 9781482471182 (library bound) | ISBN 9781482471199 (ebook)
Subjects: LCSH: Fennec–Juvenile literature.
Classification: LCC QL737.C22 R35 2026 | DDC 599.775–dc23

Published in 2026 by
Gareth Stevens Publishing
2544 Clinton Street
Buffalo, NY 14224

Designer: Jeffrey Taylor
Editor: Kristen Nelson

Photo credits: Series art (tan background) Elena_Design/Shutterstock.com; Cover, p. 1 imageBROKER.com/Shutterstock.com; p. 5 Artush/Shutterstock.com; p. 7 MUNIK KIM/Shutterstock.com; p. 9 Muhammadphotoes/Shutterstock.com; p. 11 YK/Shutterstock.com; p. 13 icemanphotos/Shutterstock.com; p. 15 bc_cornelia/Shutterstock.com; p. 17 Knihaman/Shutterstock.com; p. 19 Vine.Photographic/Shutterstock.com; p. 21 MONGKOL CHAROENSRI/Shutterstock.com.

Printed in the United States of America

CPSIA compliance information: Batch #CS26GS: For further information contact Gareth Stevens at 1-800-542-2595.

CONTENTS

Boldface words appear in the glossary.

Small and Cute

Fennec foxes are so cute! They are the smallest kind of fox on Earth. A fully grown fennec fox only weighs about 2 to 3 pounds (1 to 1.3 kg). That's about as much as a small bag of apples!

Fennec foxes are known for their big ears. These **mammals** have sand-colored fur. They stand 9 to 16 inches (23 to 41 cm) tall. Their long, fluffy tail can be about 1 foot (30 cm) long too!

Where They Live

Fennec foxes make their homes in North Africa and the Sahara. The Sahara is the largest hot **desert** on Earth! These foxes make dens in the sand dunes, or hills. They hide from **predators** in these dens.

Fennec foxes also use their den to stay out of the hot desert sun during the day. In fact, fennec foxes are mostly nocturnal, or active at night. They start to come out of their den, or burrow, when the sun starts to set.

Staying Cool

Fennec fox bodies are well **adapted** to stay cool in their desert home. Their big ears help them get rid of extra heat from their body. Their feet are furry. This **protects** their feet from the hot sand.

Fennec Fox Families

Fennec foxes live in groups of about 10. Males and females pair up for life. Mother foxes give birth to two to five kits, or babies, at a time.

Making Noise

Fennec foxes can talk to each other. They may whine, whimper, or bark, much like a dog. They let other foxes know where their **territory** is by peeing around the outside of it!

Fox Food

Fennec foxes are **omnivores**. They hunt bugs, lizards, and eggs. They also eat plant roots and fruit. They don't need to drink much water. They get water from the plants they eat.

Pet Fox?

Fennec foxes are so wildly cute, you might think one would make a good pet! But, there are laws against this in most places in the United States. Besides, since they are nocturnal, they would keep you up all night! Check one out at the zoo instead.

GLOSSARY

adapt: To change to suit conditions.

desert: A biome where it rains very little.

mammal: A warm-blooded animal that has a backbone and hair, breathes air, and feeds milk to its young.

omnivore: An animal that eats both meat and plants.

predator: An animal that hunts other animals for food.

protect: To keep safe.

territory: An area of land that an animal considers to be its own and will fight to defend.

FOR MORE INFORMATION

BOOKS

Maloney, Brenna. *Desert.* New York, NY: Children's Press, 2024.

Rose, Rachel. *The 10 Cutest Animals.* Minneapolis, MN: Bearport Publishing Company, 2025.

WEBSITES

Desert Habitiat
kids.nationalgeographic.com/nature/habitats/article/desert
Read more about life in the desert.

Science Trek
pbs.org/video/desert-habitat-d4k-aa6oni/
Watch a video about the plants and animals that live in the desert.

INDEX